D0947544

THE ORIGIN AND DEVELOPMENT OF THE QUANTUM THEORY TRANSLATED BY QUANTUM THEORY

by

Max Planck

Published by Forgotten Books 2013

Originally published 1922

PIBN 1000001241

www.ForgottenBooks.org

Free App Download

Enjoy over

1,000,000 Books

wherever you go

www.ForgottenBooks.org/apps

THE ORIGIN AND DEVELOPMENT

OF THE

QUANTUM THEORY

BY

MAX PLANCK

TRANSLATED BY

H. T. CLARKE AND L. SILBERSTEIN

BEING THE

NOBEL PRIZE ADDRESS

DELIVERED BEFORE

THE ROYAL SWEDISH ACADEMY OF SCIENCES

AT STOCKHOLM, 2 JUNE, 1920

OXFORD

AT THE CLARENDON PRESS

1922

THE ORIGIN AND DEVELOPMENT

OF THE

QUANTUM THEORY

BY

MAX PLANCK

TRANSLATED BY

H. T. CLARKE AND L. SILBERSTEIN

BEING THE

NOBEL PRIZE ADDRESS

DELIVERED BEFORE

THE ROYAL SWEDISH ACADEMY OF SCIENCES

AT STOCKHOLM, 2 JUNE, 1920

OXFORD

AT THE CLARENDON PRESS

1922

DE

OXFORD UNIVERSITY PRESS
London   Edinburgh   Glasgow   Copenhagen
New York   Toronto   Melbourne   Cape Town
Bombay   Calcutta   Madras   Shanghai
HUMPHREY MILFORD
Publisher to the University

THE ORIGIN AND DEVELOPMENT OF THE QUANTUM THEORY

My task this day is to present an address dealing with the subjects of my publications. I feel I can best discharge this duty, the significance of which is deeply impressed upon me by my debt of gratitude to the generous founder of this Institute, by attempting to sketch in outline the history of the origin of the Quantum Theory and to give a brief account of the development of this theory and its influence on the Physics of the present day.

When I recall the days of twenty years ago, when the conception of the physical quantum of 'action' was first beginning to disentangle itself from the surrounding mass of available experimental facts, and when I look back upon the long and tortuous road which finally led to its disclosure, this development strikes me at times as a new illustration of Goethe's saying, that 'man errs, so long as he is striving'. And all the mental effort of an assiduous investigator must indeed appear vain and hopeless, if he does not occasionally run across striking facts which form incontrovertible proof of the truth he seeks, and show him that after all he has moved at least one step nearer to his objective. The pursuit of a goal, the brightness of which is undimmed by initial failure, is an indispensable condition, though by no means a guarantee, of final success.

In my own case such a goal has been for many years the solution of the question of the distribution of energy in the normal spectrum of radiant heat. The discovery by Gustav Kirchhoff that the quality of the heat radiation produced in an enclosure surrounded by any emitting or absorbing bodies whatsoever, all at the same temperature. is entirely independent of the nature of such bodies (1)[1], established the existence of a universal function, which depends only upon the temperature and the wavelength. and is entirely independent of the particular properties of the substance. And the discovery of this remarkable function promised a deeper insight into the relation between energy and temperature, which is the principal problem of thermodynamics and therefore also of the entire field of molecular physics. The only road to this function was to search among all the different bodies occurring in nature, to select one of which the emissive and absorptive powers were known, and to calculate the energy distribution in the heat radiation in equilibrium with that body. This distribution should then, according to Kirchhoff's law, be independent of the nature of the body.

A most suitable body for this purpose seemed H. Hertz's rectilinear oscillator (dipole) whose laws of emission for a given frequency he had just then fully developed (2). If a number of such oscillators be distributed in an enclosure surrounded by reflecting walls, there would take place, in analogy with sources and resonators in the case of sound, an exchange of energy by means of the emission and reception of electro-magnetic waves, and finally what is known as black body radiation corresponding to Kirchhoff's law should establish itself in the vacuum-enclosure. I expected, in a way which certainly seems at the present day somewhat naïve, that the laws of classical electrodynamics would suffice, if one adhered sufficiently to generalities and avoided too special hypotheses, to account in the main for

[1] The numbers in brackets refer to the notes at the end of the article.

the expected phenomena and thus lead to the desired goal. I thus first developed in as general terms as possible the laws of the emission and absorption of a linear resonator, as a matter of fact by a rather circuitous route which might have been avoided had I used the electron theory which had just been put forward by H. A. Lorentz. But as I had not yet complete confidence in that theory I preferred to consider the energy radiating from and into a spherical surface of a suitably large radius drawn around the resonator. In this connexion we need to consider only processes in an absolute vacuum, the knowledge of which, however, is all that is required to draw the necessary conclusions concerning the energy changes of the resonator.

The outcome of this long series of investigations, of which some could be tested and were verified by comparison with existing observations, e. g. the measurements of V. Bjerknes (3) on damping, was the establishment of a general relation between the energy of a resonator of a definite free frequency and the energy radiation of the corresponding spectral region in the surrounding field in equilibrium with it (4). The remarkable result was obtained that this relation is independent of the nature of the resonator, and in particular of its coefficient of damping—a result which was particularly welcome, since it introduced the simplification that the energy of the radiation could be replaced by the energy of the resonator, so that a simple system of one degree of freedom could be substituted for a complicated system having many degrees of freedom.

But this result constituted only a preparatory advance towards the attack on the main problem, which now towered up in all its imposing height. The first attempt to master it failed: for my original hope that the radiation emitted by the resonator would differ in some characteristic way from the absorbed radiation, and thus afford the possibility of applying a differential equation, by the integration of which a particular condition for the composition of the stationary radiation could be reached, was not realized. The resonator reacted only to those rays which were emitted by itself, and exhibited no trace of resonance to neighbouring spectral regions.

Moreover, my suggestion that the resonator might be able to exert a one-sided, i. e. irreversible, action on the energy of the surrounding radiation field called forth the emphatic protest of Ludwig Boltzmann (5), who with his more mature experience in these questions succeeded in showing that according to the laws of the classical dynamics every one of the processes I was considering could take place in exactly the opposite sense. Thus a spherical wave emitted from a resonator when reversed shrinks in concentric spherical surfaces of continually decreasing size on to the resonator, is absorbed by it, and so permits the resonator to send out again into space the energy formerly absorbed in the direction from which it came. And although I was able to exclude such singular processes as inwardly directed spherical waves by the introduction of a special restriction, to wit the hypothesis of ' natural radiation ', yet in the course of these investigations it became more and more evident that in the chain of argument an essential link was missing which should lead to the comprehension of the nature of the entire question.

The only way out of the difficulty was to attack the problem from the opposite side, from the standpoint of thermodynamics, a domain in which I felt more at home. And as a matter of fact my previous studies on the second law of thermodynamics served me here in good stead, in that my first impulse was to bring not the temperature but the entropy of the resonator into relation with its energy, more accurately not the entropy itself but its second derivative with respect to the energy, for it is this differential coefficient that has a direct physical significance for the irreversibility of the exchange of energy between the resonator and the radiation. But as I was at that time too much devoted to pure phenomenology to inquire more closely into the relation between entropy and probability, I felt compelled to limit myself to the available experimental results. Now, at that time, in 1899, interest was centred on the law of the distribution of energy, which had not long before been proposed by W. Wien (6), the experimental verification of which had been undertaken by F. Paschen in Hanover and by O. Lummer and E. Pringsheim of the Reichsanstalt, Charlottenburg. This law expresses the intensity of radiation in terms of the temperature by means of an exponential function. On calculating the relation following from this law between the entropy and energy of a resonator the remarkable result is obtained that the reciprocal value of the above differential coefficient, which I shall here denote by R, is proportional to the energy (7). This extremely simple relation can be regarded as an adequate expression of Wien's law of the distribution of energy ; for with the dependence on the energy that of the wave-length is always directly given by the well-established displacement law of Wien (8).

Since this whole problem deals with a universal law of

nature, and since I was then, as to-day, pervaded with a view that the more general and natural a law is the simpler it is (although the question as to which formulation is to be regarded as the simpler cannot always be definitely and unambiguously decided), I believed for the time that the basis of the law of the distribution of energy could be expressed by the theorem that the value of R is proportional to the energy (9). But in view of the results of new measurements this conception soon proved untenable. For while Wien's law was completely satisfactory for small values of energy and for short waves, on the one hand it was shown by O. Lummer and E. Pringsheim that considerable deviations were obtained with longer waves (10), and on the other hand the measurements carried out by H. Rubens and F. Kurlbaum with the infra-red residual rays (*Reststrahlen*) of fluorspar and rock salt (11) disclosed a totally different, but, under certain circumstances, a very simple relation characterized by the proportionality of the value of R not to the energy but to the square of the energy. The longer the waves and the greater the energy (12) the more accurately did this relation hold.

Thus two simple limits were established by direct observation for the function R: for small energies proportionality to the energy, for large energies proportionality to the square of the energy. Nothing therefore seemed simpler than to put in the general case R equal to the sum of a term proportional to the first power and another proportional to the square of the energy, so that the first term is relevant for small energies and the second for large energies; and thus was found a new radiation formula (13) which up to the present has withstood experimental examination fairly satisfactorily. Nevertheless it cannot be regarded as having been experimentally confirmed with final accuracy, and a renewed test would be most desirable (14).

But even if this radiation formula should prove to be absolutely accurate it would after all be only an interpolation formula found by happy guesswork, and would thus leave one rather unsatisfied. I was, therefore, from the day of its origination, occupied with the task of giving it a real physical meaning, and this question led me, along Boltzmann's line of thought, to the consideration of the relation between entropy and probability; until after some weeks of the most intense work of my life clearness began to dawn upon me, and an unexpected view revealed itself in the distance.

Let me here make a small digression. Entropy, according to Boltzmann, is a measure of a physical probability, and the meaning of the second law of thermodynamics is that the more probable a state is, the more frequently will it occur in nature. Now what one measures are only the differences of entropy, and never entropy itself, and consequently one cannot speak, in a definite way, of the absolute entropy of a state. But nevertheless the introduction of an appropriately defined absolute magnitude of entropy is to be recommended, for the reason that by its help certain general laws can be formulated with great simplicity. As far as I can see the case is here the same as with energy. Energy, too, cannot itself be measured; only its differences can. In fact, the concept used by our predecessors was not energy but work, and even Ernst Mach, who devoted much attention to the law of conservation of energy but at the same time strictly avoided all speculations exceeding the limits of observation, always abstained from speaking of energy itself. Similarly in the early days of thermochemistry one was content to deal with heats of reaction, that is to say again with differences of energy, until Wilhelm Ostwald emphasized that many complicated calculations could be materially shortened if energies instead of calorimetric numbers were used. The additive constant which thus remained undetermined for energy was later finally fixed by the relativistic law of the proportionality between energy and inertia (15).

As in the case of energy, it is now possible to define an absolute value of entropy, and thus of physical probability, by fixing the additive constant so that together with the energy (or better still, the temperature) the entropy also should vanish. Such considerations led to a comparatively simple method of calculating the physical probability of a given distribution of energy in a system of resonators, which yielded precisely the same expression for entropy as that corresponding to the radiation law (16); and it gave me particular satisfaction, in compensation for the many disappointments I had encountered, to learn from Ludwig Boltzmann of his interest and entire acquiescence in my new line of reasoning.

To work out these probability considerations the knowledge of two universal constants is required, each of which has an independent meaning, so that the evaluation of these constants from the radiation law could serve as an a posteriori test whether the whole process is merely a mathematical artifice or has a true physical meaning. The first constant is of a somewhat formal nature; it is connected with the definition of temperature. If temperature were defined as the mean kinetic energy of a molecule in a perfect gas, which is a minute energy indeed, this constant would have the value $\frac{2}{3}$ (17). But in the conventional scale of temperature the constant assumes (instead of $\frac{2}{3}$) an extremely small value, which naturally is intimately connected with the energy of a single molecule, so that its accurate determination would lead to the calculation of the mass of a molecule and of associated magnitudes. This constant is frequently termed Boltzmann's constant, although to the best of my knowledge Boltzmann himself never introduced it (an odd circumstance, which no doubt can be explained by the fact that he, as appears from certain of his statements (18), never believed it would be possible to determine this constant accurately). Nothing can better illustrate the rapid progress of experimental physics within the last twenty years than the fact that during this period not only one, but a host of methods have been discovered by means of which the mass of a single molecule can be measured with almost the same accuracy as that of a planet.

While at the time when I carried out this calculation on the basis of the radiation law an exact test of the value thus obtained was quite impossible, and one could scarcely hope to do more than test the admissibility of its order of magnitude, it was not long before E. Rutherford and H. Geiger (19) succeeded, by means of a direct count of the a-particles, in determining the value of the electrical elementary charge as $4 \cdot 65 . 10^{-10}$, the agreement of which with my value $4 \cdot 69 . 10^{-10}$ could be regarded as a decisive confirmation of my theory. Since then further methods have been developed by E. Regener, R. A. Millikan, and others (20), which have led to a but slightly higher value.

Much less simple than that of the first was the interpretation of the second universal constant of the radiation law, which, as the product of energy and time (amounting on a first calculation to $6 \cdot 55 . 10^{-27}$ erg. sec.) I called the elementary quantum of action. While this constant was absolutely indispensable to the attainment of a correct expression for entropy—for only with its aid could be determined the magnitude of the 'elementary region' or 'range' of probability, necessary for the statistical treatment of the problem (21)—it obstinately withstood all attempts at fitting it, in any suitable form, into the frame of the classical theory. So long as it could be regarded as infinitely small, that is to say for large values of energy or long periods of time, all went well; but in the general case a difficulty arose at some point or other, which became the more pronounced the weaker and the more rapid the oscillations. The failure of all attempts to bridge this gap soon placed one before the dilemma: either the quantum of action was only a fictitious magnitude, and, therefore, the entire deduction from the radiation law was illusory and a mere juggling with formulae, or there is at the bottom of this method of deriving the radiation law some true physical concept. If the latter were the case, the quantum would have to play a fundamental role in physics, heralding the advent of a new state of things, destined, perhaps, to transform completely our physical concepts which since the introduction of the infinitesimal calculus by Leibniz and Newton have been founded upon the assumption of the continuity of all causal chains of events.

Experience has decided for the second alternative. But that the decision should come so soon and so unhesitatingly was due not to the examination of the law of distribution of the energy of heat radiation, still less to my special deduction of this law, but to the steady progress of the work of those investigators who have applied the concept of the quantum of action to their researches.

The first advance in this field was made by A. Einstein, who on the one hand pointed out that the introduction of the quanta of energy associated with the quantum of action seemed capable of explaining readily a series of remarkable properties of light action discovered experimentally, such as Stokes's rule, the emission of electrons, and the ionization of gases (22), and on the other hand, by the identification of the expression for the energy of a system of resonators with the energy of a solid body, derived a formula for the specific heat of solid bodies which on the whole represented it correctly as a function of temperature, more especially exhibiting its decrease with falling temperature (23). A number of questions were thus thrown out in different directions, of which the accurate and many-sided investigations yielded in the course of time much valuable material. It is not my task to-day to give an even approximately complete report of the successful work achieved in this field ; suffice it to give the most important and characteristic phase of the progress of the new doctrine.

First, as to thermal and chemical processes. With regard to specific heat of solid bodies, Einstein's view, which rests on the assumption of a single free period of the atoms, was extended by M. Born and Th. von Karman to the case which corresponds better to reality, viz. that of several free periods (24) ; while P. Debye, by a bold simplification of the assumptions as to the nature of the free periods, succeeded in developing a comparatively simple formula for the specific heat of solid bodies (25) which excellently represents its values, especially those for low temperatures obtained by W. Nernst and his pupils, and which, moreover, is compatible with the elastic and optical properties of such bodies. But the influence of the quanta asserts itself also in the case of the specific heat of gases. At the very outset it was pointed out by W. Nernst (26) that to the energy quantum of vibration must correspond an energy quantum of rotation, and it was therefore to be expected that the rotational energy of gas molecules would also vanish at low temperatures. This conclusion was confirmed by measurements, due to A. Eucken, of the specific heat of hydrogen (27); and if the calculations of A. Einstein and O. Stern, P. Ehrenfest, and others have not as yet yielded completely satisfactory agreement, this no doubt is due to our imperfect knowledge of the structure of the hydrogen atom. That 'quantized' rotations of gas molecules (i. e. satisfying the quantum condition) do actually occur in nature can no longer be doubted, thanks to the work on absorption bands in the infra-red of N. Bjerrum, E. v. Bahr, H. Rubens and G. Hettner, and others, although a completely exhaustive explanation of their remarkable rotation spectra is still outstanding.

Since all affinity properties of a substance are ultimately determined by its entropy, the quantic calculation of entropy also gives access to all problems of chemical affinity. The absolute value of the entropy of a gas is characterized by Nernst's chemical constant, which was calculated by O. Sackur by a straightforward combinatorial process similar to that applied to the case of the oscillators (28), while H. Tetrode, holding more closely to experimental data, determined, by a consideration of the process of vaporization, the difference of entropy between a substance and its vapour (29).

While the cases thus far considered have dealt with states of thermodynamical equilibrium, for which the measurements could yield only statistical averages for large numbers of particles and for comparatively long periods of time, the observation of the collisions of electrons leads directly to the dynamic details of the processes in question. Therefore the determination, carried out by J. Franck and G. Hertz, of the so-called resonance potential or the critical velocity which an electron impinging upon a neutral atom must have in order to cause it to emit a quantum of light, provides a most direct method for the measurement of the quantum of action (30). Similar methods leading to perfectly consistent results can also be developed for the excitation of the characteristic X-ray radiation discovered by C. G. Barkla, as can be judged from the experiments of D. L. Webster, E. Wagner, and others.

The inverse of the process of producing light quanta by the impact of electrons is the emission of electrons on exposure to light-rays, or X-rays, and here, too, the energy quanta following from the action quantum and the vibration period play a characteristic role, as was early recognized from the striking fact that the velocity of the emitted electrons depends not upon the intensity (31) but only on the colour of the impinging light (32). But quantitatively also the relations to the light quantum, pointed out by Einstein (p. 13), have proved successful in every direction, as was shown especially by R. A. Millikan, by measurements of the velocities of emission of electrons (33), while the importance of the light quantum in inducing photochemical reactions was disclosed by E. Warburg (34).

Although the results I have hitherto quoted from the most diverse chapters of physics, taken in their totality, form an overwhelming proof of the existence of the quantum of action, the quantum hypothesis received its strongest support from the theory of the structure of atoms (Quantum Theory of Spectra) proposed and developed by Niels Bohr. For it was the lot of this theory to find the long-sought key to the gates of the wonderland of spectroscopy which since the discovery of spectrum analysis up to our days had stubbornly refused to yield. And the way once clear, a stream of new knowledge poured in a sudden flood, not only over this entire field but into the adjacent territories of physics and chemistry. Its first brilliant success was the derivation of Balmer's formula for the spectrum series of hydrogen and helium, together with the reduction of the universal constant of Rydberg to known magnitudes (35); and even the small differences of the Rydberg constant for these two gases appeared as a necessary consequence of the slight wobbling of the massive atomic nucleus (accompanying the motion of electrons around it). As a sequel came the investigation of other series in the visual and especially the X-ray spectrum aided by Ritz's resourceful combination principle, which only now was recognized in its fundamental significance.

But whoever may have still felt inclined, even in the face of this almost overwhelming agreement—all the more convincing, in view of the extreme accuracy of spectroscopic measurements—to believe it to be a coincidence, must have been compelled to give up his last doubt when A. Sommerfeld deduced, by a logical extension of the laws of the distribution of quanta in systems with several degrees of freedom, and by a consideration of the variability of inert mass required by the principle of relativity, that magic formula before which the spectra of both hydrogen and helium revealed the mystery of their 'fine structure' (36), as far as this could be disclosed by the most delicate measurements possible up to the present, those of F. Paschen (37)—a success equal to the famous discovery of the planet Neptune, the presence and orbit of which were calculated by Leverrier [and Adams] before man ever set eyes upon it. Progressing along the same road, P. Epstein achieved a complete explanation of the Stark effect of the electrical splitting of spectral lines (38), P. Debye obtained a simple interpretation of the K-series (39) of the X-ray spectrum investigated by Manne Siegbahn, and then followed a long series of further researches which illuminated with greater or less success the dark secret of atomic structure.

After all these results, for the complete exposition of which many famous names would here have to be mentioned, there must remain for an observer, who does not choose to pass over the facts, no other conclusion than that the quantum of action, which in every one of the many and most diverse processes has always the same value, namely $6 \cdot 52 \cdot 10^{-27}$ erg. sec. (40), deserves to be definitely incorporated into the system of the universal physical constants. It must certainly appear a strange coincidence that at just the same time as the idea of general relativity arose and scored its first great successes, nature revealed, precisely in a place where it was the least to be expected, an absolute and strictly unalterable unit, by means of which the amount of action contained in a space-time element can be expressed by a perfectly definite number, and thus is deprived of its former relative character.

Of course the mere introduction of the quantum of action does not yet mean that a true Quantum Theory has been established. Nay, the path which research has yet to cover to reach that goal is perhaps not less long than that from the discovery of the velocity of light by Olaf Romer to the foundation of Maxwell's theory of light. The difficulties which the introduction of the quantum of action into the well-established classical theory has encountered from the outset have already been indicated. They have gradually increased rather than diminished; and although research in its forward march has in the meantime passed over some of them, the remaining gaps in the theory are the more distressing to the conscientious theoretical physicist. In fact, what in Bohr's theory served as the basis of the laws of action consists of certain hypotheses which a generation ago would doubtless have been flatly rejected by every physicist. That with the atom certain quantized orbits [i.e. picked out on the quantum principle] should play a special rôle could well be granted; somewhat less easy to accept is the further assumption that the electrons moving on these curvilinear orbits, and therefore accelerated, radiate no energy. But that the sharply defined frequency of an emitted light quantum should be different from the frequency of the emitting electron would be regarded by a theoretician who had grown up in the classical school as monstrous and almost inconceivable.

But numbers decide, and in consequence the tables have been turned. While originally it was a question of fitting in with as little strain as possible a new and strange element into an existing system which was generally regarded as settled, the intruder, after having won an assured position, now has assumed the offensive; and it now appears certain that it is about to blow up the old system at some point. The only question now is, at what point and to what extent this will happen. If I may express at the present time a conjecture as to the probable outcome of this desperate struggle, everything appears to indicate that out of the classical theory the great principles of thermo-dynamics will not only maintain intact their central position in the quantum theory, but will perhaps even extend their influence. The significant part played in the origin of the classical thermodynamics by mental experiments is now taken over in the quantum theory by P. Ehrenfest's hypo-thesis of the adiabatic invariance (41); and just as the principle introduced by R. Clausius, that any two states of a material system are mutually interconvertible on suitable treatment by reversible processes, formed the basis for the measurement of entropy, just so do the new ideas of Bohr show a way into the midst of the wonderland he has discovered.

There is one particular question the answer to which will, in my opinion, lead to an extensive elucidation of the entire problem. What happens to the energy of a light-quantum after its emission? Does it pass outwards in all directions, according to Huygens's wave theory, continually increasing in volume and tending towards infinite dilution? Or does it, as in Newton's emanation theory, fly like a pro-jectile in one direction only? In the former case the quantum would never again be in a position to concentrate its energy at a spot strongly enough to detach an electron from its atom; while in the latter case it would be neces-sary to sacrifice the chief triumph of Maxwell's theory—the continuity between the static and the dynamic fields—and with it the classical theory of the interference phenomena which accounted for all their details, both alternatives leading to consequences very disagreeable to the modern theoretical physicist.

Whatever the answer to this question, there can be no doubt that science will some day master the dilemma, and what may now appear to us unsatisfactory will appear from a higher standpoint as endowed with a particular harmony and simplicity. But until this goal is reached the problem of the quantum of action will not cease to stimulate research, and the greater the difficulties encountered in its solution the greater will be its significance for the broadening and deepening of all our physical knowledge.

NOTES

The references to the literature are not claimed to be in any way complete, and are intended to serve only for a preliminary orientation.

(1) G. Kirchhoff, Über das Verhaltnis zwischen dem Emissionsvermogen und dem Absorptionsvermogen der Korper für Warme und Licht. *Gesammelte Abhandlungen.* Leipzig, J. A. Barth, 1882, p. 597 (§ 17).

(2) H. Hertz, *Ann. d. Phys.* **36**, p. 1, 1889.

(3) *Sitz.-Ber. d. Preuss. Akad. d. Wiss.* Febr. 20, 1896. *Ann. d. Phys.* **60**, p. 577, 1897.

(4) *Sitz.-Ber. d. Preuss. Akad. d. Wiss.* May 18, 1899, p. 455.

(5) L. Boltzmann, *Sitz.-Ber. d. Preuss. Akad. d. Wiss.* March 3, 1898, p. 182.

(6) W. Wien, *Ann. d. Phys.* **58**, p. 662, 1896.

(7) According to Wien's law of the distribution of energy the dependence of the energy U of the resonator upon the temperature is given by a relation of the form :

$$U = a \cdot e^{-\frac{\beta}{T}}.$$

Since

$$\frac{1}{T} = \frac{dS}{dU},$$

where S is the entropy of the resonator, we have for R as used in the text:

$$\frac{d^2 S}{dU^2} = -\frac{\alpha}{U}$$

(8) According to Wien's displacement law, the energy U of the resonator with the natural vibration period ν is expressed by :

(9) *Ann. d. Phys.* **1**, p. 719, 1900.

(10) O. Lummer und E. Pringsheim, *Verhandl. der Deutschen Physikal. Ges.*, **2**, p. 163, 1900.

(11) H. Rubens and F. Kurlbaum, *Sitz.-Ber. der Preuss. Akad d. Wiss.* Oct. 25, 1900, p. 929.

(12) It follows from the experiments of H. Rubens and F. Kurlbaum that, for high temperatures, $U = cT$. Then, in accordance with the method quoted in (7):

(13) Put

then by integration,

$$\frac{1}{\tau} \quad dS \quad 1, \quad (. \quad bc)$$

whence the radiation formula,

$$U = bc : (e^{-b/T} - 1).$$

Cf. *Verhandlungen der Deutschen Phys. Ges.* Oct. 19, 1900, p. 202.

(14) Cf. W. Nernst und Th. Wulf, *Verh. d. Deutsch. Phys. Ges.* **21**, p. 294, 1919.

(15) For the absolute value of the energy is equal to the product of the inert mass and the square of light velocity.

(16) *Verhandlungen der Deutschen Phys. Ges.* Dec. 14, 1900, p. 237.

(17) Generally, if k be the first radiation constant, the mean kinetic energy of a gas molecule is:

If we put, therefore, $T = U$, then $k = \frac{2}{3}$. In the conventional [absolute Kelvinian] temperature scale, however, T is defined by putting the temperature difference between boiling and freezing water equal to 100.

(18) Cf. for example L. Boltzmann, Zur Erinnerung an Josef Loschmidt, *Populäre Schriften*, p. 245, 1905.

(19) E. Rutherford and H. Geiger, *Proc. Roy. Soc.* A. Vol. **81**, p. 162, 1908.

(20) Cf. R. A. Millikan, *Phys. Zeitschr.* **14**, p. 796, 1913.

(21) The evaluation of the probability of a physical state is based upon counting that finite number of equally probable special cases by which the corresponding state is realized; and in order sharply to distinguish these cases from one another, a definite concept of each special case has necessarily to be introduced.

(22) A. Einstein, *Ann. d. Phys.* **17**, p. 132, 1905.

(23) A. Einstein, *Ann. d. Phys.* **22**, p. 180, 1907.

(24) M. Born und Th. v. Karman, *Phys. Zeitschr.* **14**, p. 15, 1913.

(25) P. Debye, *Ann. d. Phys.* **39**, p. 789, 1912.

(26) W. Nernst, *Phys. Zeitschr.* **13**, p. 1064, 1912.

(27) A. Eucken, *Sitz.-Ber. d. preuss. Akad. d. Wiss.* p. 141, 1912.

(28) O. Sackur, *Ann. d. Phys.* **36**, p. 958, 1911.

(29) H. Tetrode, *Proc. Acad. Sci. Amsterdam*, Febr. 27 and March 27, 1915.

(30) J. Franck und G. Hertz, *Verh. d. Deutsch. Phys. Ges.* **16**, p. 512, 1914.

(31) Ph. Lenard, *Ann. d. Phys.* **8**, p. 149, 1902.

(32) E. Ladenburg, *Verh. d. Deutschen Phys. Ges.* **9**, p. 504, 1907.

(33) R. A. Millikan, *Phys. Zeitschr.* **17**, p. 217, 1916.

(34) E. Warburg, Über den Energieumsatz bei photochemischen Vorgangen in Gasen. *Sitz.-Ber. d. preuss. Akad. d. Wiss.* from 1911 onwards.

(35) N. Bohr, *Phil. Mag.* **30**, p. 394, 1915.

(36) A. Sommerfeld, *Ann. d. Phys.* **51**, pp. 1, 125, 1916.

(37) F. Paschen, *Ann. d. Phys.* **50**, p. 901, 1916.

(38) P. Epstein, *Ann. d. Phys.* **50**, p. 489, 1916.

'(39) P. Debye, *Phys. Zeitschr.* **18**, p. 276, 1917.

(40) E. Wagner, *Ann. d. Phys.* **57**, p. 467, 1918.

(41) P. Ehrenfest, *Ann. d. Phys.* **51**, p. 327, 1916.

AT THE CLARENDON PRESS

SPACE AND TIME in Contemporary Physics. An Introduction to the Theory of Relativity and Gravitation, by MORITZ SCHLICK. Rendered into English by HENRY L. BROSE, with an Introduction by F. LINDEMANN. 1920. 8vo (9 × 6), pp. xii + 90. 6s. 6d. net.

A GENERAL COURSE OF PURE MATHEMATICS, from Indices to Solid Analytical Geometry, by A. L. BOWLEY. 1913. In nine sections: Algebra; Geometry; Trigonometry; Explicit Functions, Graphic Representation, Equations; Limits; Plane Co-ordinate Geometry; Differential and Integral Calculus: Imaginary and Complex Quantities; Co-ordinate Geometry in Three Dimensions. With ninety-seven figures and answers to Examples. 8vo (9 × 6), pp. xii + 272. 7s. 6d. net.

A TREATISE ON STATICS WITH APPLICATIONS TO PHYSICS, by G. M. MINCHIN. Two volumes. 8vo (9 × 6).
Vol. I. Equilibrium of Coplanar Forces. Seventh edition, 1915. With 245 figures and 389 examples (300 being new to this edition). Pp. 476. 12s. net.
Vol. II. Non-Coplanar Forces. Fifth edition, revised by H. T. GERRANS, 1915. With sixty-one figures, and examples 90–389 of Vol. I + 555. Pp. 378. 12s. net.

ALGEBRA OF QUANTICS, an introduction, having as its primary object the explanation of the leading principles of invariant algebra, by E. B. ELLIOTT. Second edition, with additions. 1913. 8vo (9 × 6), pp. xvi + 416. 16s. net.

PLANE ALGEBRAIC CURVES, by HAROLD HILTON. 1920. 8vo (9 × 6), pp. xvi + 388. 28s. net.

THEORY OF GROUPS OF FINITE ORDER, an Introduction which aims at introducing the reader to more advanced treatises and original papers on Groups of finite order. By H. HILTON. 1908. With numerous examples and hints for the solution of the most difficult. 8vo (9 × 6), pp. xii + 236. 14s. net.

HOMOGENEOUS LINEAR SUBSTITUTIONS. A collection for the benefit of the Mathematical Student of those properties of the Homogeneous Linear Substitution with real or complex co-efficients of which frequent use is made in the Theory of Groups and the Theory of Bilinear Forms and Invariant Factors. By H. HILTON. 1914. 8vo (9¼ × 6), pp. viii + 184. 12s. 6d. net.

INTRODUCTION TO THE INFINITESIMAL CALCULUS, with applications to Mechanics and Physics, presenting the ffundamental principles and applications of the Differential and Integral Calculus in as simple a form as possible. By G. W. CAUNT. 1914. 8vo (9¼ × 6¼), pp. xx + 568. 12s. 6d. net.

INTRODUCTION TO ALGEBRAICAL GEOMETRY, by A. CLEMENT JONES. 1912. For the beginner and candidate for a Mathematical Scholarship, the syllabus for the Honour School of the First Public Examination at Oxford being taken as a maximum limit. With answers and index. 8vo (9¼ × 6), pp. 548. 12s. 6d. net.

CREMONA'S ELEMENTS OF PROJECTIVE GEOMETRY, translated by C. LEUDESDORF. Third edition. 1913. With 251 figures. 8vo (8¾ × 6¼), pp. xx + 304. 15s. net.

CREMONA'S GRAPHICAL STATICS, being two treatises on the Graphical Calculus and Reciprocal Figures, translated by T. H. BEARE. 1890. With 143 figures. 8vo (9 × 6), pp. xvi + 162. 8s. 6d. net.

ON THE TRAVERSING OF GEOMETRICAL FIGURES, by J. COOK WILSON. In three parts: Analytical Method, Constructive Method, and Application of the Principle of Duality. 1905. 8vo (9 × 6), pp. 166, with Addendum. 6s. 6d. net.

A TREATISE ON THE CIRCLE AND THE SPHERE, by J. L. COOLIDGE. 1916. 8vo (9 × 6), pp. 604, with thirty figures. 25s. net.

NON-EUCLIDEAN GEOMETRY, the Elements, by J. L. COOLIDGE. 1909. 8vo (9 × 6), pp. 292. 16s. net.

41313146R00022

Made in the USA
Lexington, KY
07 May 2015